FLORA OF TROPICAL EAST AFRICA

CACTACEAE

D. R. Hunt

Highly specialized fleshy perennials of diverse habit. Stems terete, globular, flattened or fluted, mostly leafless and variously spiny; spines always arising from complex axillary structures (areoles). Flowers solitary or rarely clustered, sessile (except in *Pereskia*), almost always bisexual, usually regular. Perianth segments ∞, closely imbricated in bud, in a sepaloid to petaloid series, ± free or fused below to form a short or elongate tube (hypanthium). Stamens ∞; filaments variously inserted on or at base of perianth; anthers 2-thecous, splitting longitudinally. Ovary almost always inferior, with 3–∞ parietal placentas; ovules ∞; style single; stigma-lobes 3–∞. Fruit a dry or juicy berry, often spiny, bristly or scaly. Seeds ∞, variously arillate or carunculate, with or without endosperm; embryo straight or curved.

A large, almost exclusively American family, several of whose members are grown as ornamentals in East Africa. The epiphytic genus *Rhipsalis* is sometimes considered to be indigenous in Africa, and a number of *Opuntia* species have become widely naturalized.

Terrestrial shrubs with showy yellow, orange or red flowers . . . . 1. **Opuntia**
Epiphytes with small cream or white flowers . . . 2. **Rhipsalis**

1. OPUNTIA

Mill., Gard. Dict., Abr. ed. 4 (1754)

Nopalea Salm-Dyck, Cact. Hort. Dyck., ed. 2: 63 (1850)

Austrocylindropuntia Backeb., Blätter für Kakteenforschung 1938, pt. 6 (1938)

Shrubby or arborescent cacti with cylindrical, club-shaped, subglobose or flattened branches. Areoles tufted with barbed bristles (glochids) and usually 1–∞ longer stouter spines. Leaves small, subulate to terete, usually early deciduous. Flowers solitary, sessile; perianth rotate or erect, brightly coloured, segments ∞. Ovules circinotropous (the funicles circinnate). Berry bearing areoles, glochids, and sometimes spines. Seeds ∞, encased by the hard, white funicular aril; endosperm scanty.

A large genus distributed naturally throughout the Americas from British Columbia to the Strait of Magellan. Widely introduced in the warmer parts of the world, some species as food-plants for the cochineal insect, some for forage or for their edible fruits.

Key to naturalized species

1. Joints (branch segments) cylindric, never flattened; areoles spiny, seated on pronounced tubercles and subtended at first by a fleshy terete leaf up to 4–7 cm. long; flowers brownish-red . . . . . 1. *O. exaltata*
   Joints mostly flattened or compressed, not obviously tuberculate 2

2. Areoles of mature ultimate joints with 1-several acicular spines 3
 Areoles usually spineless, occasionally with 1 or more short thin spines 4
3. Spines 1–2 per areole on the ultimate joints, the longer up to 7 cm. long; perianth yellow, usually streaked or shaded red . . 3. *O. vulgaris*
 Spines usually 3–5 per areole, the longest ± 2 cm. long; perianth yellow or orange . . 5. *O. sp. A*
4. Perianth yellow, longer than the style and stamens; joints greyish-green . . . 2. *O. ficus-indica*
 Perianth red, sheathing the exserted elongate style and stamens; joints green . . . 4. *O. cochenillifera*

1. **O. exaltata** *A. Berger*, Hort. Mortol.: 410 (1912); Britton & Rose, The Cactaceae 1: 76 (1919). Described from living plants in Sir Thomas Hanbury's garden at La Mortola, Italy

A shrub or tree up to 5 m. tall, with main stems up to 10(–30) cm. in diameter. Ultimate joints cylindrical, often curved, 3–4 cm. in diameter; areoles on raised tubercles; glochids brown or wanting; spines 1–3(–5) per areole (–12 on old wood), yellow or brownish, up to 3–7 cm. long (–13 cm. on old wood). Leaves fleshy, terete, 1–5(–7) cm. long. Flowers 8 cm. long, 4 cm. in diameter; perianth dark red to orange. Fruit pear-shaped, green.

KENYA. Naivasha, cultivated, 2 Nov. 1961, *Reid* in *E.A.H.* H291/61!; Nairobi, cultivated in Corydon Museum Ground, 5 July 1962, *Lucas* in *E.A.H.* 12592!
DISTR. **K**3, 4; South America. Introduced into East Africa around Nairobi where it is said to have escaped and to grow wild.

SYN. *Austrocylindropuntia exaltata* (A. Berger) Backeb. in Cactac. Jahrb. Deutsch. Kakt. Ges. 1941, pt. 2: 13 (1942) in obs.

NOTE. This species is close to *O. subulata* (Muehlenpf.) Engelm., also native to South America, but was regarded by Britton & Rose as " probably distinct ". One of the specimens cited above (*Lucas* in *E.A.H.* 12592) is reported as bright green, the colour attributed to *O. subulata*, not the grey-green of typical *O. exaltata*. In leaf and spine characters the specimens are better matched with *O. exaltata*.

2. **O. ficus-indica** (*L.*) *Mill.*, Gard. Dict., ed. 8 (1768); Britton & Rose, The Cactaceae 1: 177 (1919). Based on a plant in Leiden Botanic Garden, origin unknown

Shrubby or arborescent, up to 4(–5) m. tall, often with a cylindrical trunk. Joints elliptic to narrowly obovate, flattened, often 30–40 cm. long, 15–20 cm. broad, 1–1·5 cm. thick, greyish-green; glochids yellow, deciduous; spines usually none, sometimes 1 or more, up to 1·5 cm. long, bristle-like. Leaves subulate, 3–4 mm. long, early deciduous. Flowers 5–8 cm. in diameter; perianth spreading, yellow or orange, longer than the style and stamens. Ovary cylindrical, 3·5–5 cm. long, with many areoles, the upper bearing bristles up to 1·5 cm. long. Fruit ellipsoidal or obovoid, 5–9 cm. long, 3–6 cm. in diameter, variable in colour, with edible pulp.

KENYA. Naivasha, 23 Sept. 1962, *Lucas* 278!; Masai District: Narok, Gwaso Nyiro, 8 Aug. 1961, *Glover, Gwynne, Samwell* [*Paulo*] *& Tucker* 2375!
DISTR. **K**3, 6; presumably native to America but not known wild there; widely cultivated and naturalized around the Mediterranean Sea, the Red Sea, in Africa and Australia. Cultivated in Niarobi (Dec. 1948, *Bally* 6558 in *C.M.* 13559!) and reported as very common and spreading rapidly around Naivasha (*Lucas* 278)
HAB. Roadsides and by rivers; grasslands on pumice dust

SYN. *Cactus ficus-indicus* L., Sp. Pl.: 468 (1753)
 Several species and cultivars have been distinguished on minor spine and fruit characters.

3. **O. vulgaris** *Mill.*, Gard. Dict., ed. 8 (1768); Britton & Rose, The Cactaceae 1: 156 (1919). Based on an illustration by Bauhin (Hist. Pl. 1 : 154 (1650)), after L'Obel (Icones 2: 241 (1591)), origin unknown

Erect or sprawling shrub up to 3–4 m. tall, sometimes with a definite cylindric trunk. Joints usually narrowly obovate, flattened, 10–30 cm. long, 5–12 cm. broad, bright green; areoles densely and shortly woolly; glochids brownish; spines 1–2 per areole (–12 on trunk), stout acicular, up to 7(–10) cm. long, 1·5 mm. in diameter, one generally conspicuously longer than the other when present, greyish, dark-tipped. Leaves subulate, 2–3 mm. long, early deciduous. Flowers 5–7·5 cm. in diameter; outer perianth-segments yellow, striped or shaded red, inner golden yellow, widely spreading. Fruit obovoid, 5–7·5 cm. long, 2–4 cm. in diameter, glochidiate, spineless, reddish-purple, with edible pulp. Fig. 1, p. 4.

KENYA. Kiambu District: Dagoretti, 29 Jan. 1960, *Verdcourt* 2623!; Nairobi, 25 Oct. 1945, *Bally* in *C.M.* 12377!; Kwale District: Taru, 30 Apr. 1962, *Lucas, Jeffrey & Kirrika* 268!

TANGANYIKA. Pangani, 20 Jan. 1937, *Greenway* 4862!; Rufiji District: Mafia I., Kilindoni, Oct. 1937, *Greenway* 5400!

ZANZIBAR. Zanzibar I., Marahubi, Mar. 1930, *Vaughan* 1361! Naturalized along beach at Mbweni, according to U.O.P.Z.: 385 (1949)

DISTR. **K**4, 6, 7; **T**1, 3, 6; **Z**; native of South America; widely introduced elsewhere, notably in South Africa and India where it was commonly used in the cochineal dye industry

HAB. Bushland, coastal bluffs, roadsides; often planted to make fences; 0–1650 m.

SYN. *Cactus opuntia* L., Sp. Pl.: 468 (1753) pro parte minore, quoad syn. Bauhin
[*Opuntia dillenii* sensu R. O. Williams, U.O.P.Z.: 385 (1949), *non* (Ker-Gawl.) Haw.]

4. **O. cochenillifera** (*L.*) *Mill.*, Gard. Dict., ed. 8 (1768), as "*O. cochinelifera*". Based on plants figured by Dillenius, Plukenet and Sloane

Shrubby or arborescent, up to 4 m. tall, with cylindrical jointed trunks up to 20 cm. thick and branches of flattened elliptic or narrowly elliptic joints 15–40(–50) cm. long and 5–12 cm. broad, green. Glochids brown, deciduous; spines usually none, rarely one or more short spines developed on older joints. Leaves subulate, 3 mm. long, soon deciduous. Flowers ± 1·5 cm. in diameter, 5–6 cm. long from base of ovary to tips of stigmas; perianth erect, red; stamens and style exserted. Fruit red, ±5 cm. long.

TANGANYIKA. Lushoto District: S. Pare Mts., Buiko, 15 July 1942, *Greenway* 6602!
DISTR. **T**3; probably native of Jamaica and tropical America; widely cultivated
HAB. Deciduous bushland; 520 m.

SYN. *Cactus cochenillifer* L., Sp. Pl.: 468 (1753)
Nopalea cochenillifera (L.) Salm-Dyck, Cact. Hort. Dyck. 1849: 64 (1850)

NOTE. Detached stem-segments root easily and propagation by this means is commoner than by seed.

Imperfectly known species

5. **O.** (series Streptacanthae) **sp. A**

Represented by a specimen (*E.A.H.* 12526), collected in 1958 at Naivasha (**K**3), without further field notes. Possibly *O. streptacantha* Lem. or *O. megacantha* Salm-Dyck.

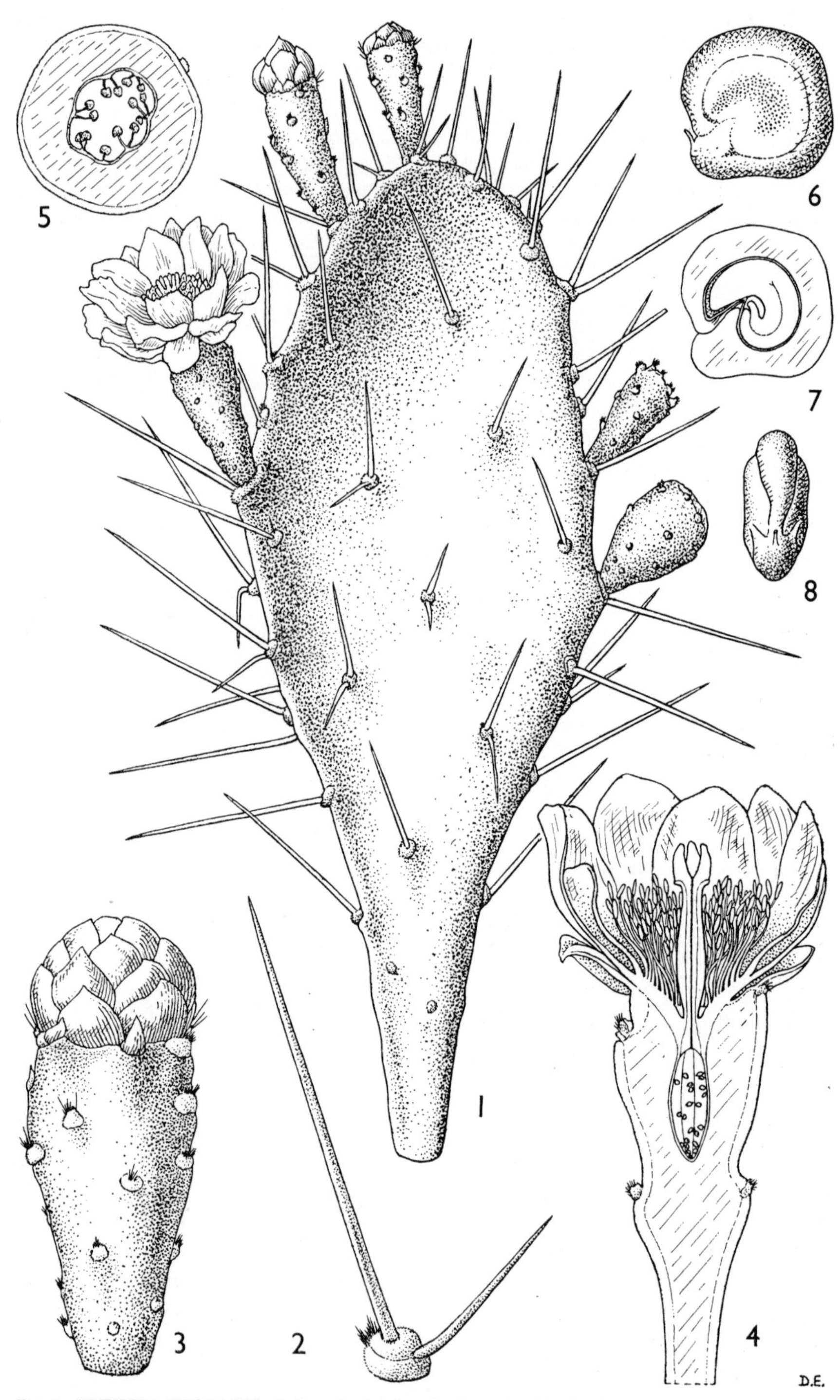

FIG. 1. *OPUNTIA VULGARIS*—**1,** flowering joint, × ½; **2,** areole with glochids and spines, × 1; **3,** flower-bud, × 1⅓; **4,** flower in longitudinal section, × 1⅓; **5,** ovary in transverse section, × ⅔; **6,** seed, side view, × 4; **7,** seed in longitudinal section, × 4; **8,** seed, end view, × 4. 1, 2, 4, from *Bally* in *C.M.* 12377; 3, from *Greenway* 4862; 5, from *Klerck & Dreyer* (South Africa); 6–8, from *du Toit* in *Nat. Herb.* 23033.

2. RHIPSALIS

Gaertn., Fruct. Sem. 1: 137 (1788), *nom. conserv.*

Hariota Adans., Fam. 2: 249 (1763)

Mostly epiphytes. Stems cylindric, angled or flattened. Areoles very small, slightly woolly, with hairs or bristles in some species. Leaves absent or reduced to minute scales. Flowers usually solitary, sessile, small; perianth-segments 5–12. Fruit white or coloured, naked or with a few scales. Seeds carunculate, with viscid endosperm.

A genus of perhaps 50 species, all confined to tropical and subtropical America except for *R. fasciculata* Haw. and *R. cereuscula* Haw. (Brazil and Madagascar) and the species below.

R. baccifera (*J. Mill.*) *W. T. Stearn* in adnot. Cact. Journ. 7: 107 (1939). Type: Represented by the description and plate, Class IX, Ord. 1 (1771) in J. Mill., Illustr. Sex. Syst. Linn. (1771–7)

Much-branched pendent epiphytic shrub 1–3(–9) m. long. Stems slender, cylindric, 2–8 mm. in diameter, sometimes slightly furrowed, arising in pairs or clusters from tips of older branches, pale green, occasionally producing adventitious roots. Areoles subtended by minute triangular scales, naked or with 1(–2) short stiff bristles 1 mm. long. Flowers lateral, solitary, small, white or yellowish. Perianth-segments 9–12, outermost triangular, inner oblong, obtuse, 2–3 mm. long. Stamens ± 10–20, inserted at base of perianth, ± as long as perianth. Style thick; stigmas 3–5. Berry spherical or somewhat elongate, 5–8 mm. in diameter, pellucid, white, pink or red. Seeds black, glossy, subpyriform, ± 1 mm. long. Fig. 2, p. 6.

UGANDA. Kigezi District: Kayonza, 23 July 1938, *Wickham* in *A. S. Thomas* 2327!; Masaka District: Nkose I., 22 Jan. 1956, *Dawkins* 863!

KENYA. Northern Frontier Province: Mt. Marsabit, July 1933, *Gardner* in *F.D.* 3216 in *C.M.* 13499!; Kericho District: SW. Mau Forest Reserve, Timbilil, Jan. 1961, *Kerfoot* 2723!; Teita Hills, 7 Feb. 1953, *Bally* 8794 in *C.M.* 20670!

TANGANYIKA. Musoma District: Mara R. Guard Post about 140 km. from Seronera, 4 Oct. 1961, *Greenway* 10234!; Moshi District: Weru Weru R., 18 Nov. 1943, *Moore* in *E.A.H.* H79/43!; Lushoto District: Amani, 19 Jan. 1950, *Verdcourt* 49!

DISTR. **U**2, 4; **K**1, 4–7; **T**1–3, 7; ? **Z**; widespread in tropical Africa from Ivory Coast east to Ethiopia and as far south as Natal; Madagascar, Mascarene Is., Ceylon; Florida, West Indies, Central and tropical South America

HAB. Epiphytic or in humus on rocks; lowland and upland rain-forest and upland evergreen bushland; 550–2200 m.

SYN. *Cassyt[h]a baccifera* J. Mill., Illustr. Sex. Syst. Linn. Class IX, Ord. I (1771)
Rhipsalis cassutha Gaertn., Fruct. Sem. 1: 137, t. 28 (1788) (*R.* "*cassytha*" auct. mult.); Oliv., F.T.A. 2: 581 (1871). Type: A specimen sent to Gaertner from Kew by Sir Joseph Banks, probably not preserved
? *R. zanzibarica* Weber in Rev. Hort. 64: 425 (1892). Described from a living plant sent to the Jardin des Plantes in Paris from Zanzibar by Sacleux
R. erythrocarpa K. Schum. in P.O.A. C: 282 (1895). Type: Tanganyika, Kilimanjaro, *Volkens* 1581 (B, holo. †, K, iso.!)
[*R. lindbergiana* sensu Rol. Goss. in Bull. Soc. Bot. Fr. 59: 100 (1912), saltem quoad syn. *R. erythrocarpam* K. Schum.]

NOTE. Seedlings and juvenile growth bear little resemblance to the adult, having short 4(–7)-angled stems with abundant areoles on the angles bearing fine bristly hairs.

Forms with pink or red berries, instead of white, have been described from East Africa (*R. erythrocarpa* K. Schum.) and tropical America (*R. cassutha* var. *rhodocarpa* Weber). This difference in fruit colour does not appear to be correlated with other characters. Nearly all the limited East African gatherings with pink or red berries are, however, from the foothills of Kilimanjaro, so that the variant may possibly be interpreted as a minor geographical race.

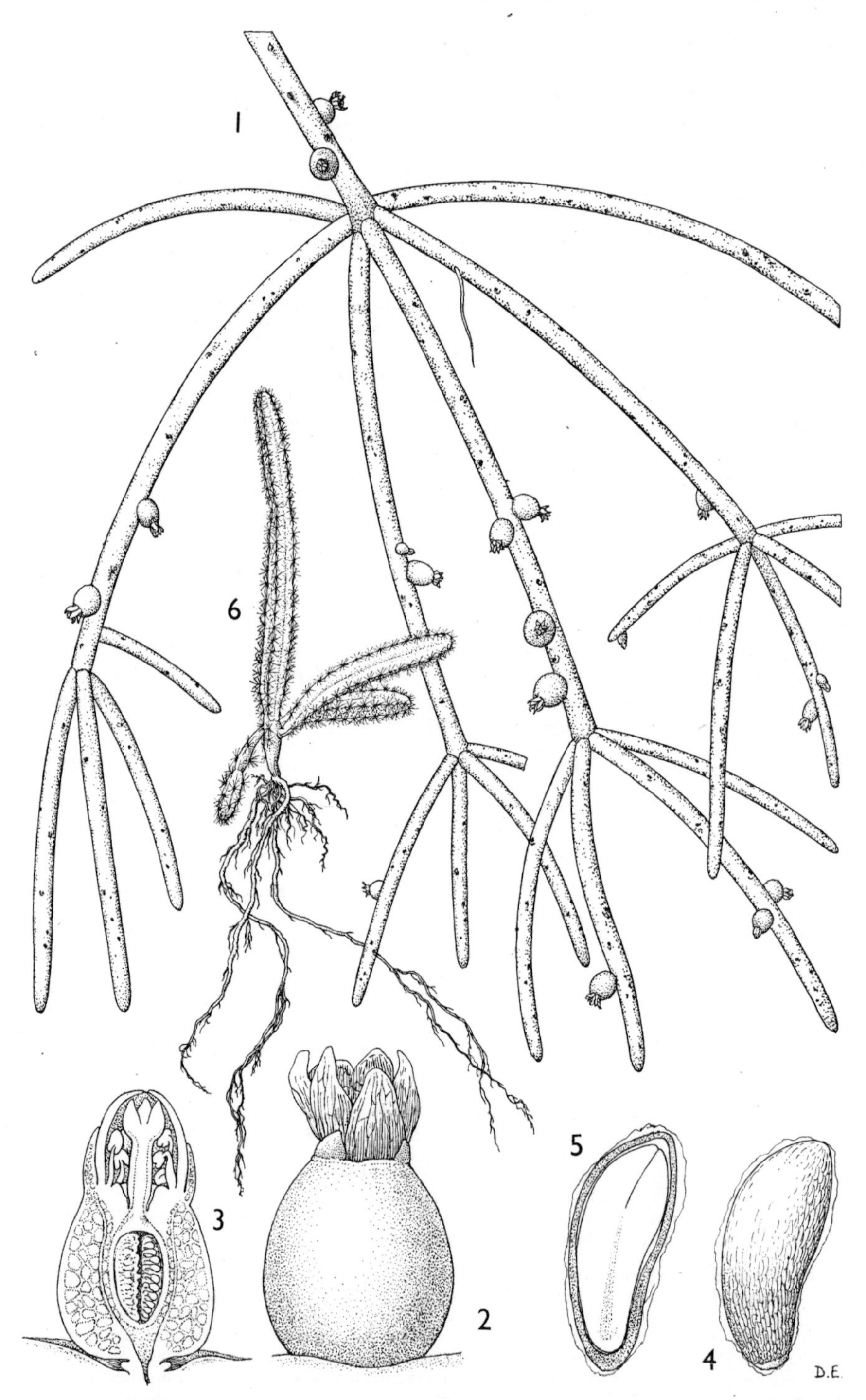

FIG. 2. *RHIPSALIS BACCIFERA*—**1,** habit, × $\frac{2}{3}$; **2,** old flower, × 6; **3,** mature bud in longitudinal section, × 6; **4,** seed, × 20; **5,** seed in longitudinal section, × 20; **6,** juvenile plant, × 1. 1, from *Gardner* in *F.D.* 3216; 2, 3, from *Greenway* 10234; 4, 5, from *Verdcourt* 49; 6, from *Verdcourt* 156.

INDEX TO CACTACEAE